A Nobel Prize Nomination: Dr. Gilbert Levin for the 1976 Discovery of Life on Mars

The Mars Viking Labeled Release Experiments

*****

Confirmation of Dr. Levin's Discovery

*****

The Conclusive Evidence of Life on Mars

*****

Respectfully Submitted to the Nobel Prize Committee by R. Gabriel Joseph, Ph.D.

Cosmology.com
Cosmology@Cosmology.com
April 16, 2018

A PDF of this document is available at:
http://Cosmology.com/NobelNominationLevin.pdf

Respectfully Submitted to the Nobel Prize Committee

A Nobel Prize Nomination: Dr. Gilbert Levin for the 1976 Discovery of Life on Mars

CONTENTS

1. Discovery, Supporting Evidence, and Visual Confirmation of Life on Mars

It takes courage to speak truth to power and challenge the "authority" of the status quo, and uncommon genius to devise experiments which lead to paradigm-shifting discoveries that revolutionize our understanding of the nature of life in the universe. Dr. Gilbert Levin is a man of uncommon vision and courage, a seeker of truth, and has the added distinction of being the first scientist to discover life on Mars--an achievement worthy of the Nobel Prize.

In 1976, NASA stated that the Mars Viking life detection experiments might be "perhaps the most important experiment in the history of science." Dr. Levin's Viking Labeled Release (LR) experiment, one of three included on Viking, was designed to detect biological activity on Mars (1-4).

In the course of developing the Viking Labeled Release (LR) experiment, and prior to transport to Mars, thousands of laboratory and field tests were performed and the LR experiment proved capable of detecting a very wide range of microorganisms including bacteria, algae and fungi.

The Viking LR experiment was elegant and straight-forward. It took a sample of Martian soil and added a nutrient containing radioactive carbon. The presence of radioactivity in the gasses released was evidence of active metabolism. A control experiment treated a second sample that had been heat-treated to kill microorganisms. In every experiment conducted, the control (heat treated) sample yielded negative results whereas positive results were obtained from the raw sample, including evidence of biological reproduction. Both Viking landing sites, some 4,000 miles apart, produced strong responses and met the pre-mission criteria for the detection of biological activity, and life on Mars (1-4).

Dr. Levin's Mars Viking LR data are supported by evidence of (A) waxing and waning and increasing and decreasing concentrations of methane at a variety of localities both at ground level and in the atmosphere of Mars (5-9), the most logical source of which is biological activity; (B) biochemical analyses and detection of biological residue in "meteors" ejected from Mars (10-15); (C) simulation studies demonstrating various organisms can survive and flourish in a Mars-like environment (16-20); (D) Martian structures and specimens with features resembling stromatolites (21-24) --the likely remnants of cyanobacteria/blue-green algae; (E) visual observation of greenish substances which change in size and appear to be algae (23-25); and (F) the expert judgment of 40 experts in fungi, algae and lichens and 30 experts in geomorphology and mineralogy, who after closely examining photos of Martian specimens taken by NASA's Rover's Curiosity and Opportunity, formed a statistically significant consensus that these were living organisms resembling terrestrial mushrooms and other species of fungi (26-27). Indeed, many of these experts referred to the visual evidence as "obvious" (26).

It is the visual evidence --photographed by NASA and presented at the end of this document-- and the consensus of 70 experts (26-27), coupled with recent evidence of biological activity (5-9) and the observations of additional investigators (23-25)-- that have conclusively confirmed Dr. Levin's 1976 discovery. Thus, the honorable members of the Nobel Committee are herewith asked to review these data, and Dr. Gilbert Levin's description of the positive findings from the Viking LR experiment, and to consider formally nominating this scientist for the Nobel Prize in Physiology or Medicine.

The Nobel Committee is also asked to consider that --despite all the evidence of current biological activity on Mars (1-15, 21-27) --NASA plans to harvest and transport Martian specimens to Earth in just a few years, while claiming there is no life on Mars (28) and thus no need for public concern or oversight by elected officials who might rightfully fear contagion, disease, plague, and may cancel NASA's plans.

By officially recognizing Dr. Levin's great discovery, and the supporting evidence, the Nobel Committee would alert the world to the potential dangers of NASA's plans to transport specimens from Mars to Earth, and in so doing, "shall have conferred the greatest benefit on mankind" as stated in Alfred Nobel's will.

2. The Viking Labeled Release Experiment Detects Life on Mars

As summarized by Levin (1), the Viking life detection experiments were based on the assumption that, if complex life forms had evolved on Mars, they would have to be accompanied by microorganisms to complete the life cycle, similar to terrestrial life, and that Martian life would be carbon-based, and that its biochemical reactions would be aqueous. In the course of developing the LR experiment, and prior to transport to Mars, thousands of laboratory and field tests were performed. The LR proved capable of detecting life in all viable samples, over a very wide range of microorganisms, including pure and mixed cultures of aerobic, anaerobic and facultative bacteria, as well as algae, fungi, lichen, and sulfur bacteria. In every case, negative LR controls verified the biological nature of the initial responses. Viability was confirmed by classical microbial methods. In those few cases where the classical method did not show life and the LR did, the control proved the LR correct and highly sensitive. Thus, a very strong case for reliability of the LR was established.

The LR experiment was therefore included in the Viking biology package, which, paradoxically, included two additional experiments -- the Gas Exchange (GEx) and the Pyrolytic Release (PR) -- which were questionable at best (1). Indeed, the GEx and the PR proved unable to detect the presence of microbes on Earth which were living in permafrost or frozen tundra. By contrast, and as described by Levin (1) "the LR approach is unique among life detection systems in that it is not based on static chemical or physical properties in the sample, but on the detection of on-going metabolism. The method is extremely sensitive."

Once on Mars, the Viking sampling arm obtained Martian soil samples which were

4

distributed to all three life detection instruments. As described by Levin (1), "The LR instruments operated flawlessly on Mars. Both Viking landing sites, some 4,000 miles apart, produced strong responses and met the pre-mission criteria for the detection of life (2-4). Thus, the LR experiment proved there was biological activity, and life on Mars."

In a further effort to distinguish between biological and non-biological agents, additional, more defining controls were executed by commands from Earth. Each such ad hoc run again demonstrated on-going Martian metabolism. Four different LR experiments were conducted, each of which yielded positive results, and five controls all of which supported the positive results as biological.

Levin concluded that the LR experiment had proved there is life on Mars, and that the "amplitudes and kinetics of the Mars LR results were similar to those of terrestrial results, especially close to those of soils in, or from, frigid areas."

3. Why NASA Rejected the Positive LR Findings

Days after the positive findings of life on Mars were announced to the world, the results were rejected by administrators and NASA scientists who were shocked and upset by the findings (29). Many believed and argued the results were contrary to Torah and the Bible and NASA shouldn't be doing these studies (20,32).

As summed up by Massachusetts Institute of Technology professor, Arthur Lafleur, who helped design the Viking Mission GC-MS: NASA administrators wanted the Viking life detection experiment to fail and so they rejected the LR results, because: "There were a lot of people at the time who couldn't handle the discovery of life on Mars... It was a philosophical thing; they just didn't want that result. And those prejudices" resulted in a rejection of the data (29). NASA administrators then discredited the evidence for life on Mars, and issued a statement that "the LR had not detected life on Mars, but had detected a chemical or physical agent that had produced false positive results" (1).

Subsequently, despite claims that detecting life on Mars might constitute one of the most important experiments in history, NASA refused to equip any subsequent mission with life detection experiments, and has harassed and threatened those who've continued the search (31).

4. The Torah, Bible and NASA vs Evidence of Extraterrestrial Life

As first detailed by Dr. Levin in the 1970s: when he met with NASA administrators Dr. John Olive (and 20 others), Dean Cowie of the Carnegie Institute, and the editor of the journal Science (Dr. Phil Abelson), some of these "scientists" began evoking the Bible as proof against life on Mars (32). As related by Dr Levin: "Abelson became angry and began shouting: "The Bible tells us there cannot be any life on other planets." Abelson and Dean Cowie then walked out of the meeting, arm-in-arm, as Abelson continued to rave about the Bible.

Dr. Levin's findings were repeatedly rejected by those in positions of authority and who claimed the "Bible" and "Torah" proves there is no life on Mars. There were hundreds of religiously devout, high ranking scientists working at NASA, including, Dr. Abe Silverstein who founded a religious Temple in Cleveland Ohio and who appointed nearly half the male congregation to positions at NASA (33); and Dr. Velvl Greene who admitted that he was warned by religious zealots that searching for life on Mars was "contrary to Torah," and he "shouldn't be doing this kind of work...because it goes contrary to Torah... It's forbidden by Jewish law." (30). Dr. Greene states he felt confused and so sought a "religious dialogue" with his Rabbi to whom he explained "what we're doing is just normal bacteriology; it's not very exciting..." His Rabbi responded: "Let me decide that." As stated by Dr. Greene: "I... assembled a pile of unclassified documents--three or four thick folders--and I sent them all to the Rebbe" who then reviewed this evidence and may have contributed to the decision about life on Mars.

Is evidence of extraterrestrial life contrary to religious faith? Torah and the Bible also does not mention Jupiter. Certainly this does not mean Jupiter does not exist. However, it is this "reasoning" --based on religious (32) and philosophical objections (29) -- that contributed to the rejection of any and all evidence of life on Mars.

Dr. Levin was not the first scientist to fall victim to religious extremism at NASA. Drs. Greene, Silverstein, and other members of the Hebrew and Christian faith, had been appointed to positions of high authority at NASA in 1960 (30,33). In 1961 and over the following two years, three brilliant scientists, George Claus, Bartholomew Nagy, and Nobel Laureate Harold Urey, reported discovering biological residue and an assortment of fossilized bacteria, including algae and cyanobacteria, in meteors older than this solar system (34-38). The discovery led to world wide publicity, followed by a campaign of threats, harassment, intimidation, slander and defamation by NASA scientists and administrators who dispersed funds to successfully discredit these scientists (39) who were described as "frauds" who had "deliberately contaminated" the meteors and had perpetrated a "hoax" (39-42).

Was Harold Urey a "fraud"? Dr. Urey, whose pioneering work on isotopes won him the Noble Prize in chemistry in 1934, had worked on the development of the atomic bomb, and had continued to make amazing discoveries in the advancement of science, including in the 1950s when he and his former student, Stanley Miller, produced over 20 amino acids, the building blocks of life, in an experiment designed to determine whether some of the chemical conditions of Earth early in its history could initiate the genesis of life (43).

However, when Claus, Nagy, and Urey published their findings proving the existence of fossilized extraterrestrial life, NASA launched, directed, and funded a vicious campaign of slander and denial and in actions reminiscent of the trials of Galileo, NASA repeatedly demanded that these scientists recant and retract their findings, which, at first, they refused to do. Finally, their reputations destroyed and under pressure from NASA, these scientist agreed to change the description of their findings.

NASA's bullying campaign to discredit any and all findings indicative of extraterrestrial life, have continued into the present. Case in point, the Journal of Cosmology (JOC) had been founded in 2009, with an editorial policy of "openness to all ideas even those the editors disagree with," so long as the article is scholarly, backed up by scientific references, and survives an objective peer review. In less than 2 years, JOC had become famous in the United States, having published the works of over 1,000 scientists from almost every major university in the English-speaking world--including 40 scientists from NASA--with editions edited by Sir Roger Penrose of Oxford University, and two editions edited by NASA Senior Scientists (31, 43), including Dr. Joel Levine who held a press conference at NASA headquarters in February of 2011, praising JOC's exemplary editorial policies of peer review (31).

However, two months later, in May of 2011, when NASA scientist, Richard Hoover, published evidence of extraterrestrial microfossils in JOC (44) --and which generated world-wide attention-- NASA administrators verbally assaulted Hoover and demanded he retract his findings (31). Hoover refused. NASA's Chief Scientist, Paul Hertz then directed NASA's wrath at JOC which he and others at NASA defamed, slandered, and libeled; referring to JOC as a "joke", "not a real journal" and falsely claiming JOC "does not peer review" (31). In so doing, these religious zealots destroyed the reputations of JOC and Richard Hoover. In fact (as reported by Science magazine in 2011) it was most likely NASA's Paul Hertz who "suggested hanging Hoover in effigy in the conference center lobby" for reporting evidence of extraterrestrial microfossils (45).

By defaming and destroying JOC and threatening to murder Hoover, NASA's religious fanatics sent a clear message to the scientific community: be afraid. For over 50 years, NASA's religious zealots have sought to terrorize the scientific community and drag us back into the dark ages of the Inquisition.

5. The Consensus of 70 Experts: "Mushrooms", "Fungi", "Lichens", "Puffballs" On Mars

The author of this Nobel Prize nomination (R. Gabriel Joseph), began his scientific career by publishing, in peer reviewed scientific journals, a number of major scientific studies in neuroscience and the biology of behavior in the late 1970s (e.g. neuroplasticity and recovery of function in the primate brain; the hormonal basis of sex differences in cognition and behavior; early environmental influences on learning, memory and intelligence and synaptic development (46-52); research which continued over the following decades.

Beginning in 2005, upon examining photos taken by NASA's Mars Rover Opportunity, and later, the Rover Curiosity, this scientist observed specimens that resembled algae and "mushrooms." By 2014, this scientist had collected over 100 photos of Martian specimens which clearly resembled algae, lichens, mushrooms; and some of which, over a period of days, grew up out of the Martian soil, took mushroom shapes, and shed spores (25). Experts consulted by this scientist identified many of these specimens by name, and called the evidence of life on Mars "obvious." However, mindful of NASA's death threats directed at Hoover and the destruction of JOC's reputation, none of these scientists were willing to come forward; as they feared the consequences.

To overcome these fears, this scientist, using advanced computerized technology, devised a study in which the world's experts could examine, rate, and identify these specimens via links to a secure study-website which had been coded and programmed for exactly that purpose; but which linked their judgements to their computer's IP address, and not their names (27). Moreover, this unique system enabled experts to type in the names of the specimens, with all answers directly linked to that scientist's unique computer IP address.

It is well established that the use of computerized information technology is an excellent method of obtaining accurate, valid and reliable information as to opinions and judgments, particularly when they are from homogeneous "closed populations" (53-55). Samples from a "closed population" are generally considered to have a very high degree of validity and reliability, and which accurately represent the views of other members of those homogenous populations.

Hence, the Life on Mars study (27) --launched in 2016-- was based on a homogeneous "closed population" of 1,000 geologists who were identified by their universities as experts in mineralogy and geomorphology, and a "closed population" of 1,000 biologists identified by their universities as experts in fungi, algae and lichens. All were provided secure links (one for biologists, another for geologists) to a secure website with 25 photos of Martian specimens photographed by NASA depicting organisms previously judged to resemble fungi (25). Out of these two closed homogenous population of experts, a total of 70 scientists--30 Geologists and 40 Biologists--completed the invitation-only online study which was closed and completed after 3 days so as to maintain the integrity of the study and guard against outside influences. Given that these scientists were drawn from two homogenous populations of experts, they thus represent these experts in general and the general expertise of these two closed populations thereby conferring generalized validity and reliability (53-55).

This study was designed so each expert could determine, via a 4 point probability scale, on the likelihood these specimens are living organisms:

1 (0% Probability) - 2 (33% Probability) - 3 (66% Probability) - 4 (100% Probability).

Upon examining photographic evidence these 70 experts formed a highly statistically significant consensus that there is life on Mars; and dozens of fungal-experts identified these specimens as "puffballs,""Basidiomycota", "lichens" and "mushrooms", some of which grow out of the ground and shed spores (26-27).

Two independent monitors (Dr. R.R. University of British Columbia; Dr. H.G. George

Mason University), consulted with this scientist and provided critical feedback throughout all phases of this study, and they, and NASA's Planetary Protection Officer and NASA's Director of Astrobiology, were provided the original email lists, copies of the email invitations with email addresses; as well as all the raw data linked to each unique IP addresses, immediately after the study was completed.

A Fisher's exact test (57) was performed (by Y.A. of UCLA). Highly significant results were obtained, proving that Geologists and Biologists agreed there is a high probability of life on Mars as based on the comparisons for the top 5 pictures chosen by Biologists ($p = <0.0008$) and Geologists ($p = <0.0004$); and the same is true of the top 7 photos; Biologists ($p = <0.0001$); Geologists ($p = <0.0001$). Dozens of experts also identified these living specimens as "puff balls," "Basidiomycota" and "mushrooms" (18-19). Moreover, geologists and biologists agreed on 5 of the top 7 specimens.

Typically, the alpha level is set at < 0.05 (5%). However, in this study, the findings were significant well beneath the < 0.001 level. Hence, if one were to continuously redraw a sample from the same two populations of 1,000 biologists with an expertise in fungi, lichens, and algae, and 1,000 geologists with an expertise in geomorphology and mineralogy, we would expect to obtain the same exact results over 99.9% of the time.

The overall pattern of results confirms that fungi have colonized the Red Planet and that there is "obvious" evidence of life on Mars. This visual evidence, photographed by NASA, is presented at the conclusion of this nominating document.

The results from the 2016 study--and thus the consensus of 70 experts--coupled with recent evidence of biological activity (5-9) and the observations of additional investigators (23-25)-- confirms the findings of Dr. Gilbert Levin: There is life on Mars.

6. Religion vs Evidence of Life on Mars

At the same time the Life on Mars study (25, 27) was being planned and then carried out, and in defiance and violation of the First Amendment of the Bill of Rights, of the United States Constitution (58), NASA awarded over one million dollars U.S. to a Christian religious group "The Center of Theological Inquiry"-- for the explicit purpose of advising NASA on how to respond to reports of extraterrestrial life. As is well documented and part of the public record, NASA also attempted to prevent the 2016 Life on Mars study from going forward and this author was subject to harassment and threats by NASA personnel (31).

History is replete with examples of the religious authorities persecuting scientists, and philosophers, for arguing in favor of, or providing evidence of extraterrestrial life. The religious authorities believe this evidence is "contrary to Torah", "contrary to the Bible" and contrary to their beliefs and must be discredited. The arrest, torture and murder of Giordano Bruno, is one of the more notorious examples of the religious wars against evidence or science-based arguments favoring extraterrestrial life; battles which continue to this day.

What does the Torah and Bible actually say as related to this issue?

> "In the Beginning...
> the Spirit of God hovered over the waters..." --Genesis 1.
> What is the "Spirit of God" if not life?
> and Mars at one time had oceans of water
> --and where there is water, there is life.

It's been said: "Religion is the science of worshipping god." Science is not incompatible with belief in God. The discovery and confirmation of life on Mars is scientific evidence of the majesty, power, and glory of God whose living spirit hovered over the waters.

"And when you look up to the sky and behold the sun and the moon and the stars, the whole heavenly host, you must not be lured into bowing down to them or serving them. These the Lord your God allotted to other peoples." --Deuteronomy 4.19

Torah, the Bible, says: life, the spirit of god, is everywhere.

7. Biologically Produced Martian Methane

On Earth, 90% of all methane released into the atmosphere is produced biologically by living and decaying organisms; released as a waste product by bacteria and archaea, as well as by certain species of fungi (59). Hence, it can be assumed that 90% of methane produced on Mars also has a biological source.

In 2003, Europe's Mars Express spacecraft tracked three separate methane plumes rising into the Martian atmosphere consisting of 19,000 metric tons of methane gas (5,6). It has since been determined that Martian atmospheric methane levels vary over time and are punctuated by transient and major spikes in concentration (6-9). For example, in July of 2013, "an upper limit of 2.7 parts per billion of methane" in the general vicinity of the Gale Crater was reported whereas on September of 2013, methane levels significantly declined, fluctuating between a value of 0.18 ppbv to 1.3 ppbv (7). This was followed by a "tenfold spike" in levels of methane in the Martian atmosphere with increases in late 2013 and early 2014, averaging "7 parts of methane per billion in the atmosphere" (7-9).

It is highly unlikely, in fact, it is improbable that the cyclic, waxing, waning, waxing of Martian methane is produced by non-biological sources. On Mars, whatever seeps into the soil, rock, or which comes to be locked beneath the surface of the planet, remains locked in place, including geologically produced methane. On Mars methane is readily destroyed by chemical reactions and the remainder is blown into space due to the solar wind, the lack of a magnetosphere, and the sun's UV rays. Thus, the fact that Martian methane has been repeatedly replenished, waxing and waning then waxing in concentration, indicates a living source.Biology and the activity of Martian organisms is the only reasonable scientific explanation for the continual secretion and replenishing of methane on Mars. Thus this data must also be considered as confirmation of Dr. Levin's 1976 discovery of life on Mars (4) and is entirely consistent with the visual evidence of the presence and growth of Martian fungi and algae (21-27).

8. NASA's Dangerous Plan To Transport Martian Organisms to Earth

Dr. Alfred Nobel stated in his will that he wished to honor scientists who "shall have conferred the greatest benefit on mankind." NASA --despite all the evidence of Martian life and current biological activity on Mars (1-15, 21-27) --plans to harvest and transport Martian specimens to Earth in just a few years, while simultaneously claiming "... it is highly unlikely that living organisms will be found on the samples" (28) and thus no need for public concern or oversight by elected officials who will rightfully fear contagion, disease, plague, and may cancel NASA's plans.

These Martian samples will become the most valuable substances on Earth and they will be stolen as they arrive, just as NASA's Inspector General determined that "NASA has been experiencing loss of astromaterials since lunar samples were first returned by Apollo missions" (60) --crimes NASA administrators were party too and attempted to cover up. Lunar rocks were stolen as they arrived on Earth from the first mission to the moon --thefts that continued into the present. However, in contrast to moon rocks, these extremely valuable Martian samples will include bacteria, fungi, and possibly unknown pathogens and they too will be stolen as they arrive thus exposing all of Earth to contamination and the possibility of contagion, disease, plague, or even a sixth mass extinction. In a worst-case scenario, NASA's plan threatens all of life on this planet.

The Nobel Prize Committee, by awarding Dr. Levin the Nobel Prize, will not only honor one of the greatest discoveries in the history of science, but will "have conferred the greatest benefit on mankind" by forcing NASA and the United States government to admit to the conclusive evidence for life on Mars, and the incredible dangers of transporting Martian organisms to Earth.

9. A Nobel Prize Nomination: Dr. Gilbert Levin for the Discovery Life on Mars

In 1976, NASA stated that the Mars Viking life detection experiments were "perhaps the most important experiment in the history of science." In every experiment performed on Mars, at two landings sites 4,000 miles away, the LR results met pre-mission criteria for the detection of biological activity, and life on Mars (1-4). NASA rejected the results based on religious and philosophical concerns which had nothing to do with science, and refused to equip any subsequent mission with life detecting experiments. In fact, even before the Viking Mission was launched, in a meeting with 20 NASA administrators and the editor of the journal Science, vehement objections against the LR experiment were voiced based on the Bible (32). Dr. Levin's 1976 discovery of life on Mars, is also supported by reports from a number of independent investigators (23-26), including waxing and waning and decreases and increases in ground-level and atmospheric methane (5-9) and visual evidence of Martian specimens which closely resemble fungi--some of which grow from the surface, take mushroom shapes, and shed spores (27)--as depicted in the photos at the end of the document. Dr. Levin's results, therefore, have now been confirmed.

Having played a role in the independent verification of Levin's experimental results, as documented by the statistically significant consensus of 70 scientists identified by their universities as experts --some of whom identified these specimens by name-- it is my responsibility and my duty as a scientist to request that the Nobel Committee examine Dr. Levin's original findings, and the conclusive visual, and related evidence, and, if the Committee believes his discovery worthy, and the dangers of the Mars Return Sample Program to be of concern, to then formally nominate Dr. Gilbert Levin for the Noble Prize for the discovery of life on Mars.

Respectfully Submitted to the Nobel Committee

/s/ Rhawn Gabriel Joseph, Ph.D.
Cosmology.com
Cosmology@Cosmology.com
4/16/2018

A PDF of this document is available at:
http://Cosmology.com/NobelNominationLevin.pdf

10. References

1. Levin, G. (2010).Extant Life on Mars: Resolving the Issues, Journal of Cosmology, 5, 920-929.

2. Levin, G. V. and P. A. Straat (1976) Labeled Release - An Experiment in Radiorespirometry" Origins of Life, 7, 293-311.

3. Levin, G. V. and P. A. Straat (1979) Completion of the Viking Labeled Release Experiment on Mars, J. Mol. Evol., 14, 167-183.

4. Levin, G. V. and P. A. Straat, "The Likelihood of Methane-producing Microbes on Mars," Instruments, Methods, and Missions for Astrobiology XII, SPIE Proc., vol. 7441, invited paper 744110D, 2009.

5. Mumma, M.J., Novak, R.E., DiSanti, M.A., Bonev, B.P., (2003) A sensitive search for methane on Mars. Bull. Am. Astron. Soc. 35, 937.

6. Mumma, M.J., Villanueva, G.L., Novak, R.E., Hewagama, T., Bonev, B.P., DiSanti, M.A., Mandell, A.M., and Smith, M.D. (2009) Strong release of methane on Mars in northern summer 2003. Science. doi:10.1126/science.11,,65,243.

7. Webster, G. et al. (2013) Isotope ratios of H, C, and O in CO2 and H2O of the martian atmosphere. Science 341, 260-263.

8. Webster, G. et al. (2014) NASA Rover Finds Active and Ancient Organic Chemistry on Mars." NASA.

9. Webster, C. R. et al. (2015) Mars methane detection and variability at Gale crater, Science, 347, 415-417.

10. McKay, D.S., et al. (1996) Search for past life on Mars: possible relic biogenic activity in Martian meteorite ALH84001. Science 273: 924-930.

11. McKay, G., Mikouchi, T., Schwandt, C. & Lofgren, G. (1998) Fracture fillings in ALH84001. Feldspathic glass : carbonatic and silica. 29 the Annual Lunar and Planetary Science Conference held March 16-20, 1998 in Houston, Texas. LPI Contribution No. 1998, Abstract no. 1944.

12. McKay, D.S., Thomas-Keprta, K.L., Clemett, S.J., Gibson Jr, E.K., Spencer, L. and Wentworth, S.J. (2009) Life on Mars: new evidence from martian meteorites. In, Instruments and Methods for Astrobiology and Planetary Missions, 7441, 744102.

13. Clemett, S. J., et al (1998). Evidence for the extraterrestrial origin of polycyclic aromatic hydrocarbons in the Martian meteorite ALH84001. Faraday Discuss. 109:417-436.

14. Thomas-Keprta K.L., Clemett S.J., Bazylinski D.A., Kirschvink J.L., McKay D.S., Wentworth S.J., Vali H., Gibson E.K., Romanek C.S. (2002) Magnetofossils from Ancient Mars: A Robust Biosignature in the Martian Meteorite ALH84001. Applied and Environmental Microbiology 68, 3663-3672.

15. Thomas-Keprta, K. L., et al., (2009). Origins of magnetite nanocrystals in Martian meteorite ALH84001. Geochimica et Cosmochimica Acta, 73, 6631-6677.

16. Osman, S., Peeters, Z., La Duc, M.T., Mancinelli, R., Ehrenfreund, P., Venkateswaran, K., (2008). Effect of shadowing on survival of bacteria under conditions simulating the Martian atmosphere and UV radiation. Applied and Environmental Microbiology 74, 959-970.

17. Pacelli, C., L. Selbmann, L. Zucconi, J. P. P. De Vera, E. Rabbow, G. Horneck, R. de la Torre, S. Onofri "BIOMEX experiment: Ultrastructural alterations, molecular damage and survival of the fungus Cryomyces antarcticus after the Experiment Verification Tests." (2016) Origin of Life and Evolution of Biospheres 2016, 47(2):187-202.

18. Sanchez, F. J., E. et al. (2012) The resistance of the lichen Circinaria gyrosa (nom. provis.) towards simulated Mars conditions-a model test for the survival capacity of an eukaryotic extremophile." Planetary and Space Science, 2012, 72(1), 102-110.

19. Selbmann L, Zucconi, D. Isola, and D. Onofri (2015) Rock black fungi: excellence in the extremes. From the Antarctic to Space." 2015. Current Genetics 61: 335-345. DOI 10.1007/s00294-014-0457-7.

20. Mahaney, W. C. & Dohm, J. (2010) Life on Mars? Microbes in Mars-like Antarctic Environments, Journal of Cosmology, 2010, Vol 5, 951-958.

21. Rizzo, V., & Cantasano, N. (2009) Possible organosedimentary structures on Mars. International Journal of Astrobiology 8 (4): 267-280.

22. Bianciardi, G., Rizzo, V., Cantasano, N. (2014). Opportunity Rover's image analysis: Microbialites on Mars? International Journal of Aeronautical and Space Sciences, 15 (4) 419-433.

23. Rabb, H. (2018). Life on Mars - Visual Investigation, SoCIA, University of Nevada, Reno, USA. April 14, 2018.

24. Joseph, R (2017), Mars: Evidence of Current and Past Life; Viking LR, Meteor ALH8401, Stromatolites, Methane, Fungi. Journal of Cosmology at Cosmology.com, http://cosmology.com/MarsCurrentPastLife.html

24. Kupa, T. A. (2017). Flowing water with a photosynthetic life form in Gusav Crater on Mars, Lunar and Planetary Society, XLVIII

25. Joseph, R. (2014) Life on Mars: Lichens, Fungi, Algae, Cosmology, 22, 40-62.

26. Dass, R. S. (2017) The High Probability of Life on Mars: A Brief Review of the Evidence, Cosmology, at Cosmology.com Vol 27, April 15, 2017, http://cosmology.com/LifeOnMarsReview.html

27. Joseph, R. (2016) A High Probability of Life on Mars, the Consensus of 70 Experts, Cosmology, 25, 1-25. http://cosmology.com/LifeOnMarsStudy1.html

28. NASA, Sample Returns Collecting Rock and Soil Samples and Returning Them to Earth, https://mars.nasa.gov/programmissions/missions/missiontypes/samplereturns/

29. Schmidt, C. W. (2001) The Chemistry of Life on Mars, Chemical Innovation, Vol 31, No 5. 12-16.

30. Greene, V. (2000, 2013). Jewish Science. The Rebbe and the Scientist: Looking for Life on Mars, http://jemedia.org/email/newsletter/My_Encounter/11-9-13.pdf, https://www.chabad.org/therebbe/article_cdo/aid/2436891/jewish/The-Rebbe-and-the-Scientist-Looking-for-Life-on-Mars.htm

31. Joseph, R. vs NASA (2016). Plaintiffs Complaint for Declaratory and Injunctive Relief, Violations of the Public Trust Doctrine, Violations of First, Fifth, Ninth and Fourteenth Amendment. Filed in the Northern District Federal Court of California, CV 165142.

32. DiGregorio, B. (1997). Mars: the living planet. Frog Books.

33. Silverstein, A. The West Temple (Cleveland, Ohio) Records. 1910-2008, http://collections.americanjewisharchives.org/ms/ms0784/ms0784.html, and http://www.thewesttemple.com/about-us/our-history3

34. Claus, G., Nagy, B. (1961) A Microbiological Examination of Some Carbonaceous Chondrites. Nature 192, 594 - 596.

35. Nagy, B., Meinschein, W. G. Hennessy, D, J. (1961), Mass-spectroscopic analysis of the Orgueil meteorite: evidence for biogenic hydrocarbons. Annals of the New York Academy of Sciences 93, 25-35.

36. Nagy, B., Claus, G., Hennessy, D, J., (1962), Organic Particles embedded in Minerals in the Orgueil and Ivuna Carbonaceous Chondrites. Nature 193, 1129 - 1133.

37. Nagy, B., Fredriksson, K., Kudynowkski, J., Carlson, L. (1963), Ultra-violet Spectra of Organized Elements. Nature 200, 565 - 566.

38. Nagy, B., Fredriksson, K., Urey, C., Claus, G., Anderson, C. A., Percy, J. (1963). Electron Probe Microanalysis of Organized Elements in the Orgueil Meteorite, Nature 198, 121 - 125.

39. NASA Grant NsG-366

40. Anders, E. and F. W. Fitch (1962), Search for Organized Elements in Carbonaceous Chondrites. Science, 138, pp. 1392-1399

41. Fitch, F., H. P. Schwarcz and E. Anders (1962), 'Organized elements' in carbonaceous chondrites. Nature, 193, pp. 1123-1125

42. Fitch, F. W. and E. Anders (1963), Organized Element: Possible Identification in Orgueil Meteorite. Science, 140, pp. 1097-1100

43. Miller, S. L. & Urey, H. C. (1959). Organic Compound Synthesis on the Primitive Earth. Science. 130 (3370): 245–51.

43. Journal of Cosmology (2009-2011) http://JournalofCosmology.com; Cosmology.com

44. Hoover, R (2011). Fossils of Cyanobacteria in CI1 Carbonaceous Meteorites, Journal of Cosmology, 2011, Vol 13.

45. Science Magazine (2011), Bugs in space? Forget it. http://www.sciencemag.org/news/2011/03/bugs-space-forget-it

46. Joseph, R. and Casagrande, V. A. (1978). Visual field defects and morphological changes resulting from monocular deprivation in primates." Proceedings of the Society for Neuroscience, 1978, 4, 1982.

47. Joseph, R., Forrest, N., Fiducia, P. Como, and J. Siegel, (1980) Electrophysiological and behavioral correlates of arousal." Physiological Psychology, 1980, 9, 90-95.

48. Joseph, R. and Gallagher, R. E. (1980). Gender and early environmental influences on learning, memory, activity, overresponsiveness, and exploration." Developmental Psychobiology, 1980, 13, 527-544.

49. Casagrande V. A., and Joseph, R. (1978) Effects of monocular deprivation on geniculostriate connections in prosimian primates. Anatomical Record, 1978, 190, 359.

50. Joseph, R., Hess, S., and Birecree, E. (1978) Effects of sex hormone manipulations on exploration and sex differences." Behavioral Biology, 1978, 24, 364-377.

51. Joseph, R. and Casagrande, V. A. (1980) Visual field defects and recovery following lid closure in a prosimian primate." Behavioral Brain Research, 1980 1, 150-178.

52. Casagrande, V. A. & Joseph, R. (1980). Morphological effects of monocular deprivation and recovery on the dorsal lateral geniculate nucleus in prosimian primates. Journal of Comparative Neurology, 194, 413-426.

53. Dommeyer, C.J., P. Baum, R.W. Hanna, and K.S. Chapman. (2004). Gathering faculty teaching evaluations by in-class and online surveys: their effects on Response Rates and Evaluations. Assessment & Evaluation in Higher Education, Vol. 29, No. 5, October 2004

54. Hewson, C. and Stewart, D. W. (2016) Internet Research Methods. John Wiley & Sons. DOI:10.1002/9781118445112.stat06720.pub2

55. Richardson, J.T.E. (2005). Instruments for obtaining student feedback: a review of the literature. Assessment & Evaluation in Higher Education 30, no. 4: 387-415.

56. Watt, S., C. Simpson, C. McKillop, and V. Nunn. (2002). Electronic course surveys: does automating feedback and reporting give better results? Assessment & Evaluation in Higher Education 27, no. 4: 325-337.

57. Fisher, R. A. (1922) On the interpretation of $\chi2$ from contingency tables, and the calculation of P. Journal of the Royal Statistical Society. 85 (1): 87-94.

58. FFRF protests large NASA grant used for religious purposes, https://ffrf.org/news/news-releases/item/29063-ffrf-protests-large-nasa-grant-used-for-religious-purposes

59. U.S. Department of Energy, U.S. Department of Agriculture (2017). Complete Guide to Biogas and Methane: Agricultural Recovery, Manure Digesters, AgSTAR, Landfill Methane, Greenhouse Gas Emission Reduction and Global Methane Initiative. Complete Guide to Methane Hydrate Energy: Ice that Burns, Natural Gas Production Potential, Effect on Climate Change, Safety, and the Environment.

60. NASA Office of Inspector General, NASA (2001) Nasa's Management Of Moon Rocks And Other Astromaterials Loaned For Research, Education, And Public Display Https://Oig.Nasa.Gov/Audits/Reports/Fy12/Ig-12-007.Pdf

Figures 1-16
VISUAL EVIDENCE OF LIFE ON MARS
(vs Earth)
Consensus of 70 Experts:

Figure 1: Mushrooms From Earth which are nearly identical to Martian specimens on Mars.

Figure 2. Mars - From Joseph, R. (27). A High Probability of Life on Mars: The Consensus of 70 Experts. Photograph: NASA/Rover Opportunity.

Figure 3. Mars - Note the fluffy white spores. From Joseph, R. (27). A High Probability of Life on Mars: The Consensus of 70 Experts. Photograph: Photograph: NASA

Figure 4. Mars - From Joseph, R. (27). A High Probability of Life on Mars: The Consensus of 70 Experts. Photograph: NASA

Figure 5. Mars - From Joseph, R. (27). A High Probability of Life on Mars: The Consensus of 70 Experts. Photograph: NASA

Figure 6. Mars - From Joseph, R. (27). A High Probability of Life on Mars: The Consensus of 70 Experts. Photograph: NASA

Figure 7. - Mars From Joseph, R. (27). A High Probability of Life on Mars: The Consensus of 70 Experts. Photograph: NASA

Figure 8. Mars - From Joseph, R. (27). A High Probability of Life on Mars: The Consensus of 70 Experts. Photograph: NASA

Figure 9. Mars - From Joseph, R. (27). A High Probability of Life on Mars: The Consensus of 70 Experts. Photograph: NASA

Figure 10. Mars - Note the fluffy white spores. From Joseph, R. (27). A High Probability of Life on Mars: The Consensus of 70 Experts. Photograph: NASA

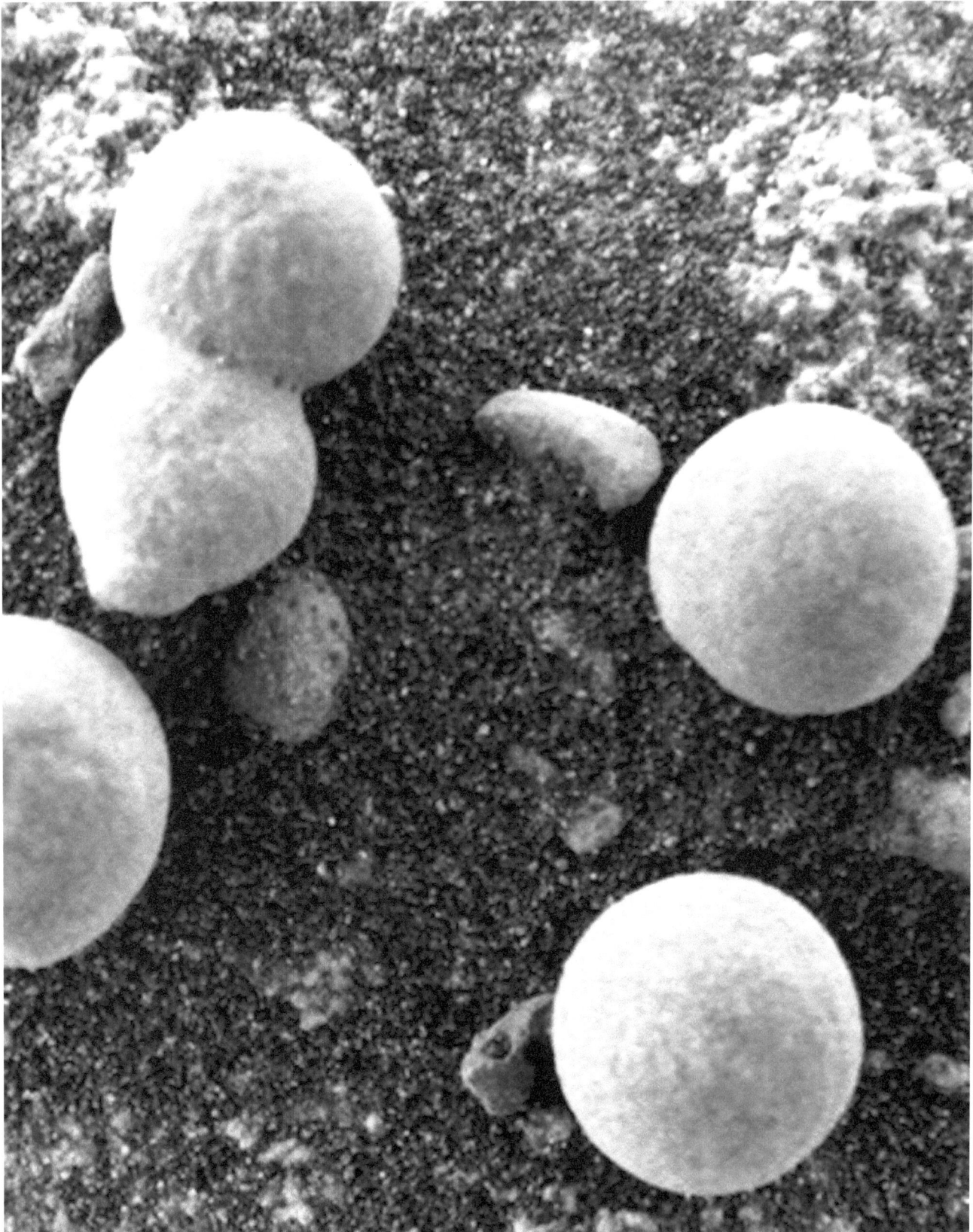

Figure 11. Mars - Note the fluffy white spores. From Joseph, R. (27). A High Probability of Life on Mars: The Consensus of 70 Experts.

Figure 12. Mars. - Sol 1145 (top) and Sol 1148 (bottom). Growth of Martian Organisms Over the Course of Day. From Joseph (27). Photograph: NASA

Figure 13 - Lichens (Earth) which are nearly identical to Martian specimens photographed on Mars by NASA's rovers.

Figure 14. Mars - From Joseph, R. (27). A High Probability of Life on Mars: The Consensus of 70 Experts. Photograph: NASA

Figure 15. Mars - (Sol 88) - From Joseph, R. (27). A High Probability of Life on Mars: The Consensus of 70 Experts. Photograph: NASA

Figure 16. Mars - From Joseph, R. (27). A High Probability of Life on Mars: The Consensus of 70 Experts. Photograph: NASA

Respectfully Submitted to the Nobel Prize Committee

A Nobel Prize Nomination: Dr. Gilbert Levin for the 1976 Discovery of Life on Mars

The Mars Viking Labeled Release Experiments

*****

Confirmation of Dr. Levin's Discovery

*****

The Conclusive Evidence of Life on Mars

/s/ Rhawn Gabriel Joseph, Ph.D.
Cosmology.com
Cosmology@Cosmology.com
4/16/2018

A PDF of this document is available at:
http://Cosmology.com/NobelNominationLevin.pdf

www.ingramcontent.com/pod-product-compliance
Lightning Source LLC
Chambersburg PA
CBHW052045190326
41520CB00002BA/192